Communications in Catering

Practical guide-lines and assignments for students in the catering industry

SHELAGH SNELL
Lecturer in Communications
Worcester Technical College

Edward Arnold

For the Catering Students of Worcester Technical College, who were the guinea-pigs.

© Shelagh Snell 1985

First published in Great Britain 1985 by
Edward Arnold (Publishers) Ltd, 41 Bedford Square, London WC1B 3DQ

Edward Arnold (Australia) Pty Ltd, 80 Waverley Road, Caulfield East, Victoria 3145, Australia

Reprinted (revised) 1987

British Cataloguing in Publication Data

Snell, Shelagh
   Communications in catering: practical guidelines and assignments for students in the catering industry.
   1. Communication in hotel management—Great Britain   2. Communication in food service management—Great Britain
   302.2   TX911.3.C6/

ISBN 0-7131-7443-9

All rights reserved. No part of this publication may be reproduced, stored in a retrieval system, or transmitted in any form or by any means, electronic, photocopying, recording, or otherwise, without the prior permission of Edward Arnold (Publishers) Ltd.

Text set in 10/12 pt Century Compugraphic by Colset Private Limited, Singapore

Printed in Great Britain by J.W. Arrowsmith Ltd., Bristol

Contents

Introduction   *iv*
Acknowledgements   *v*
Foreword   *vi*

Guide-lines on using the telephone   *1*
Guide-lines on the business letter   *4*
Assignment 1: The hotel stationery   *7*
Assignment 2: The stolen jewellery   *8*
Assignment 3: Orwell in Paris   *10*
Assignment 4: The soft-drinks survey   *12*
Assignment 5: The break-down   *14*
Guide-lines on preparing a short talk   *18*
Guide-lines on preparing a demonstration   *20*
Assignment 6: The careers evening   *21*
Assignment 7: Trouble in the kitchen   *22*
Assignment 8: The caterer's kitchen knives   *24*
Guide-lines on graphics   *25*
Assignment 9: The advertising campaign   *28*
Assignment 10: The sailing club dinner   *29*
Assignment 11: Confirmation of booking   *30*
Assignment 12: The Falcon sales conference   *31*
Assignment 13: A tourist package   *32*
Assignment 14: Love thy neighbour   *36*
Assignment 15: The hotel flowers   *38*
Guide-lines on writing a short report   *39*
Assignment 16: Consumer report   *40*
Assignment 17: Accident report   *41*
Assignment 18: A hotel in Brittany   *42*
Assignment 19: A weekend in the country   *44*
Assignment 20: The seminar   *45*
Assignment 21: Easter in Paris   *46*
Assignment 22: A visit to London   *48*
Assignment 23: The open evening   *51*
Assignment 24: How a French housewife uses chicken   *52*
Guide-lines on applying for a job   *55*
Guide-lines on the interview   *56*

Introduction

This book of guide-lines and assignments is intended to provide material for catering students following craft and BTEC courses in Hotel and Catering subjects at Colleges of Further Education, and should be very useful to students following training courses within the Catering Industry generally. I have deliberately included *oral* assignments since successful and confident oral communication within the industry is so important, and these assignments will be useful for students following ESB courses. The assignments are all extremely practical and I have tried to make them as realistic as possible. My own students are allocated, on average, an hour per week for communications; some of these assignments can be covered in one lecture period, but others will take longer to complete.

Shelagh Snell

Acknowledgements

The publishers would like to thank the following for permission to use copyright material and illustrations:

Guardian Royal Exchange (p. 9);
A M Heath, the late Sonia Brownell Orwell and Martin Secker and Warburg Ltd (pp. 10–11);
The Mansell Collection Ltd (p. 10);
Sutcliffe Design (p. 11);
Consumers' Association (pp. 13 & 37);
The Automobile Association and Ordnance Survey (pp. 14–17);
Ted Poole, The College, Swindon, (p. 20);
Letraset UK Ltd (p. 27);
Rotring (p. 26, foot);
The Econasign Co Ltd (p. 26, top);
Bristol City Council (p. 33);
British Tourist Authority (p. 35);
Department of Trade and Industry (p. 35);
Pan Books Ltd (p. 42);
The French Government Tourist Board (p. 46);
Paris Travel Service (pp. 46–7);
TrainLines of Britain for British Rail (p. 49);
London Regional Transport (p. 50);
David Higham Associates Ltd (p. 52).

I am grateful to my colleagues at Worcester Technical College for their help and encouragement in writing this book, and in particular those members of the Catering Department for whose students it was originally written.

I am also extremely grateful to Christabel Burniston, President of the English Speaking Board, for her great kindness in reading through the manuscript, and for her wisdom and advice on the oral section in particular.

Finally my thanks are due to Marjorie Honeybourne and Gill Owen who have typed my work so efficiently and cheerfully.

Foreword

It is a great pleasure to meet, at last, a book which recognises the human and personal attributes needed at all levels in the hotel and catering industry.

The efficiency and harmonious running of any such establishments depend on clear, concise and courteous communication, be it in the manager's office, at the reception desk, in the dining room or in the kitchen. The *quality* of the ever-expanding hotel and catering industry is shown in its personal service however much new technology replaces human hands.

In this book, Shelagh Snell sets out to help young men and women who are training for a variety of jobs within the industry, and from her wide experience as a Communications lecturer in F.E. she deals with the many aspects of written and oral communication which determine effective interaction.

At the request of several Colleges of Further Education the English Speaking Board have provided a welcome addition to their existing oral assessments in the form of a syllabus and certificate specifically related to hotel and catering training.

Communications in Catering sets out essential guide-lines which will help students in their preparation for these assessments, but far beyond that, it provides valuable advice and practice for all aspirants as they meet real life situations in an infinitely varied and demanding profession.

Christabel Burniston
President of the English Speaking Board

Guide-lines on using the telephone

The telephone is one of the most widely used and yet one of the most feared instruments of our daily lives. Time and time again students (and indeed many adults) have said, 'I *hate* using the telephone', with a heartfelt emphasis on the word 'hate'. Why do people hate this piece of machinery so much? They hate it because it is so impersonal, because it is impossible to see the reaction of the person at the other end of the line (sometimes a distinct advantage!) and because it is impossible to make complete contact. The only way in which we can communicate when we use the telephone is by voice and ear. We cannot use our sight and we cannot use our hands and so our voice becomes very important indeed. It is worth noting that people who use the telephone a great deal are very seldom afraid of it. Familiarity should not 'breed contempt', and 'practice' in this case really does 'make perfect'.

Remember that the telephone is the *front* of every organisation, and the person who answers the telephone is the very first contact the caller has with that organisation. So it is very important indeed to answer the telephone correctly.

Answering the telephone

The wrong way

- *Do not* pick up the receiver, breathe heavily and wait for the caller to speak first!
- *Do not* pick up the receiver and say 'Hello!' or 'Hi!' or anything casual.
- *Do not* pick up the receiver and say 'Yes?' (We have probably all experienced these faults at some time or other!)

The right way

Remember to have a pad and pencil to hand – very important!

Do pick up the receiver and say 'Good morning', 'Good afternoon', or 'Good evening' (whichever the case may be), identify your place of work, and add 'Can I help you?'

e.g. 'Good evening, *Charlie's Wine Bar*, can I help you?'

- Remember that it reassures the caller when you repeat the important parts of the call:

GOOD MORNING, RED LION HOTEL, CAN I HELP YOU?

names, addresses, times, dates.
- Try to avoid using the expression 'O.K.'. We all do it, but it's a very bad habit to get into.
- Use the caller's name when you learn it and remember to finish a call politely.

e.g. 'Thank you for calling, Mr. Smith. We look forward to seeing you on July 10th when your table will be ready for 8 o'clock. Goodbye'.

Note: 'Goodbye' *not* 'Cheerio' or 'Bye now', or anything casual!

Some situations for practice

(Two students are needed, one to make the call and one to receive it.)

1. You are taking some friends out to dinner for a birthday celebration. Ring up an hotel or restaurant and book a table.
2. An elderly relative from abroad is coming to visit you, and since you haven't a spare room in your house to accommodate her, you ring up a local hotel to book a room for her there. (Remember she is an *elderly* relative!)
3. You have the responsibility, in your hotel, for ordering the vegetables and fruit, and you have ordered a consignment of avocado pears from your wholesaler for a very important dinner. They have not arrived and you must have them by midday at the latest. Ring up the wholesaler. . .
4. You have just returned from a 'Winter Break' holiday weekend in London, and have realised you have left some ear-rings, which you value, behind in a drawer in your hotel room. Ring up the hotel. . .
5. You are going to have a party for your eighteenth birthday. Your parents have refused to have it at home and suggest that you book a room at a local hotel. Ring up the local hotel. . .
6. You have been asked by your hotel manager to ring up a Mrs. Smithson who has booked accommodation for a holiday in your hotel. Mrs Smithson is a rather exacting lady who insists on having the same room overlooking the sea every year. Unfortunately this particular room is still being decorated and is therefore unavailable. Mrs. Smithson will have to be satisfied with another room! (Have we an exacting Mrs. Smithson in the group?)
7. The president of the local St. John's Ambulance Brigade rings up your hotel to see if their annual dinner dance can be held there on a particular date. Unfortunately that date is unavailable, but you must try very hard to accommodate these valued customers.
8. You went out last evening and had a meal with some friends at a local restaurant. It was an excellent meal, but when you got home you checked the bill and found it to be incorrect. (You hadn't liked to check it in the restaurant in front of your friends.) You decide to ring the restaurant. . .
9. The secretary to the Sales Director of the Falcon Car Company telephones to inquire what facilities your hotel can offer for conferences. The secretary is obviously impressed by your reply and asks if you can accommodate 56 people for a two-day conference in March.
10. You see an advertisement in the *Hotel and Caterer* for: 'Summer staff wanted in Sea-Side Hotel. Ring Weybridge-on-Sea 5434 for further details.' You decide to ring. . .
11. A colleague has slipped in the bar of the hotel where you are both working and has cut his leg very badly. Unfortunately there isn't a car available to take him to hospital so you have to ring for an ambulance. . .
12. A guest staying at your hotel has asked you to ring a local garage to collect and repair her car which has broken down. (You must decide where the breakdown has taken place!) She needs the car for the following afternoon. Make the appropriate telephone call.

Did you remember the vital parts of the conversation – names, times, dates?

Note: *Never underestimate yourself.* As soon as you pick up a telephone you become a very important person indeed. If you are answering the telephone at your place of work, you are the first contact the caller has with your organisation, and much depends on you!

Guide-lines on the business letter

Basic rules

Layout

1 The appearance of your letter is very important indeed. The layout should always be well balanced and pleasing to look at.
2 Consider the size of the paper you're using, and 'lay' the words on it carefully, avoiding a squashed or crowded look. Make sure there is a margin all round the letter, and that there are no large gaps left at the top or bottom of the paper.
3 Take note of the personal preferences of the person for whom you are writing. Check whether he, or she, likes letters to be *fully blocked* – all the typed entries (apart from his, or her, address) to begin from the left hand margin, with no indenting of paragraphs or *semi blocked*, with the paragraphs indented and the closing of the letter centrally placed.

A good layout

```
                                14 Coach Drive
                                Longton
                                Derby
                                DC5 4CL

7 April 1987

Floral Arts Ltd
5 Chester Road
Wrexham
Clwyd

Dear Sirs

Please send me your catalogue and price list
as advertised in this week's edition of 'The
Flower Arranger'.

I would also be extremely grateful if you would
send me details of your summer courses on Floral
Decoration.

Yours faithfully



(Jane Smithson)
```

Fully-blocked style

```
                                    14, Coach Drive,
                                       Longton,
                                         Derby,
                                          DC5 4CL.

                        7th. April 1987.

Floral Arts Ltd.,
5, Chester Road,
Wrexham,
Clwyd.

Dear Sirs,

        Please send me your catalogue and price list
as advertised in this week's edition of 'The Flower
Arranger'.

        I would also be extremely grateful if you would
send me details of your summer courses on Floral
Decoration.

                        Yours faithfully,



                        (Jane Smithson)
```

Semi-blocked style

Style

1. Aim for sincerity and friendliness, but not familiarity.
2. Keep to the point, but avoid being curt.
3. Always be courteous, never 'off-hand'. Even when you reprimand someone by letter, e.g. in a letter of complaint, there is never any excuse for rudeness.
4. Make quite sure that all your facts and figures are correct. Be efficient.
5. Always use clear, simple and straightforward language. Avoid jargon and words you think sound impressive; they can often sound rather pompous.
6. If you are replying to a letter, always start by thanking the sender, e.g.

Punctuation

It has now become common practice to type business letters with the minimum of punctuation. Commas and full stops are used to make the sentences 'make sense', but other marks of punctuation such as the apostrophe, are fast disappearing. However, some people do not accept that this is time-saving (which is the intention) but see this practice as a sign of sloppiness, and a further example of the degeneration of our language; so take note of the personal preferences of the person for whom you are writing the letter. Look carefully at the following example of a semi-blocked letter which is punctuated using the traditional style.

Dear Mr. Jones,

Thank you for your letter dated June 5th...

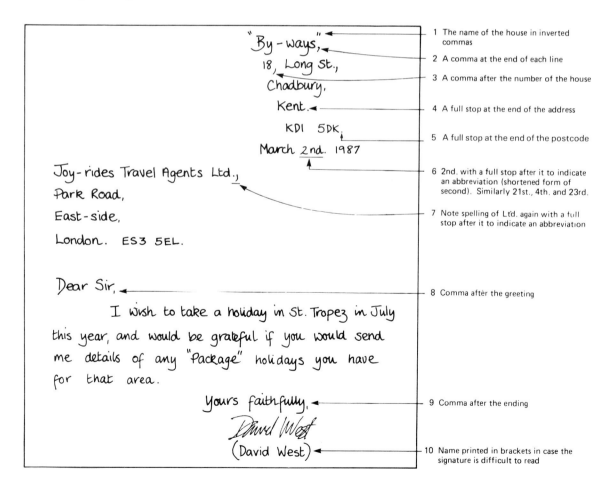

Greetings and endings

1. If you know the name of the person to whom you are writing, it is discourteous not to use it in the greeting, unless the name constitutes the name of a firm or company in which case you would use 'Dear Sir', 'Sirs', or 'Dear Madam', 'Mesdames'.
2. Use the ending 'Yours sincerely' on the following occasions:
    i) If you have met the person to whom you are writing.
    ii) If you have had previous correspondence with them.
    iii) If you have been instructed to write to a specific person.
3. The ending 'Yours faithfully' *always* goes with: Dear Sir, Sirs, Madam, Mesdames.

Examples:

1. You are answering an advertisement for a job and have been instructed to write to:
    Jane Archdale,
    Personnel Officer,
    Quality Foods Ltd.,
    9, High Street,
    Brixham,
    Devon. DV14 6BH.
    You have never met Jane Archdale before, but you have been asked to write to her specifically, so your letter will begin: *Dear Ms. Archdale*, and end: *Yours sincerely,*
2. You are writing to your previous headmaster (James Black) for a reference. Quite clearly you know the person to whom you are writing, so start your letter: *Dear Mr. Black,* and end it: *Yours sincerely,*
3. You are writing to:
    Kitchen Equipment and Design,
    4, Beech Lane,
    Highbury,
    Berks. HG4 5BK.
    to order some kitchen equipment. Kitchen Equipment and Design is the name of a firm or company, so you will write: *Dear Sirs*, and end your letter: *Yours faithfully,*
4. You are writing to:
    James Smith, Esq., ARIBA,
    Chartered Architect,
    5, College Street,
    Bath,
    Somerset. BA7 SM3.
    about the designing of a restaurant for which you are a consultant. James Smith constitutes the name of a firm (in this case an architect's practice) so you will start your letter: *Dear Sir*, and end it: *Yours faithfully*, for the initial letter at least.

Notes

1. *Esq.* after the name is short for the word *Esquire*, an historical title of rank, and is no longer used as frequently as it once was. It is, however, considered a courtesy by many people who still use it. *Esq.* is always placed after the name and before the qualifications (as shown with James Smith, above) since a man is assumed to be a gentleman before he can become an Associate of the Royal Institute of British Architects!

2. The second letter in a sequence of correspondence usually carries the ending *Yours sincerely*, since the ice has been broken and you are now deemed to know the person to whom you are writing, if only slightly.

3. *Yours* always starts with a capital Y, followed by a small *f* or *s* for *faithfully* or *sincerely*.

Assignment 1: The hotel stationery

During an idle few minutes behind the reception desk at the *Racecourse Hotel*, Worcester, you are reading a copy of *Hotel and Caterer* when you notice the following advertisement:

Give Your Hotel A New Image

Top quality stationery at very low prices.

Order your headed paper, menus and brochures now and take advantage of our special introductory price offer.

Write for catalogue and price list to:
Hotel Supplies Ltd., 9, Wareham Street, London. SE 4.

Task 1

The manager asks you to write to Hotel Supplies Ltd., for a catalogue and price list.

Task 2

Hotel Supplies Ltd. send you the catalogue but not the price list. Write an appropriate letter of complaint.

Task 3

You are the secretary to the sales manager of Hotel Supplies Ltd. Write a letter of apology to the manager of the *Racecourse Hotel*, Worcester.

Assignment 2: The stolen jewellery

You are the hotel receptionist. A guest staying at your hotel reports the loss, presumably by theft, of a valuable piece of jewellery from her room.

Task 1

Either by writing, or by direct approach, see if you can obtain a Theft Claim form from an insurance company and complete it. (Otherwise use the one illustrated as a guide and design your own.)

Before you post the form to the insurance company, you and the hotel housekeeper decide to have a last look in the guest's room (the guest has now left the hotel) and you find the piece of jewellery down the side of an easy chair.

Task 2

Write an appropriate letter to the guest to accompany the piece of jewellery which you are returning.

GRE is a member of the Insurance Ombudsman Bureau   Office of Issue

Guardian Royal Exchange Assurance

CLAIM FORM FOR FIRE THEFT MONEY AND ALL RISKS POLICIES

Policy No. _____ Branch or Agent to whom you paid your last premium _____

Name of Insured _____ Tel. No. { Home _____
Address (Private) _____ { Business _____
_____ Postcode _____
Address (Business) _____ Postcode _____
Trade or Occupation (if more than one state all) _____
Situation of premises or place where loss or damage occurred _____
_____
Date of loss or damage _____ Time _____ a.m./p.m.
Explain fully how the loss or damage occurred _____

ADDITIONAL QUESTIONS FOR THEFT MONEY AND ALL RISKS CLAIMS.

When was the loss or damage discovered?   Date _____ Time _____ a.m./p.m.
By whom was the discovery made? _____
When was the property last seen?   Date _____ Time _____ a.m./p.m.
By whom was it last seen? _____
When were the police notified? _____ Address of Police Station _____
Have any other steps been taken to recover the property? _____

PLEASE ANSWER THE FOLLOWING QUESTIONS IF THE CLAIM IS IN RESPECT OF A THEFT AT YOUR OWN PREMISES

Total value of contents of premises at time of theft £ _____ Are the premises, or any part, let or sub-let? _____
How many nights have the premises been unoccupied during the past year? _____
Was anyone in the premises at the time of the theft? _____ If so, please give names and addresses _____
_____

What steps have you or are you taking to prevent a recurrence? _____

ZC33 (7/83)   P T O

Assignment 3: Orwell in Paris

Read the following passage carefully.

Except for about an hour, I was at work from seven in the morning till a quarter past nine at night; first at washing crockery, then at scrubbing the tables and floors of the employees' dining-room, then at polishing glasses and knives, then at fetching meals, then at washing crockery again, then at fetching more meals and washing more crockery. It was easy work, and I got on well with it except when I went to the kitchen to fetch meals. The kitchen was like nothing I had ever seen or imagined – a stifling, low-ceilinged inferno of a cellar, red-lit from the fires, and deafening with oaths and the clanging of pots and pans. It was so hot that all the metal-work except the stoves had to be covered with cloth. In the middle were furnaces, where twelve cooks skipped to and fro, their faces dripping sweat in spite of their white caps. Round that were counters where a mob of waiters and *plongeurs* clamoured with trays. Scullions, naked to the waist, were stoking the fires and scouring huge copper saucepans with sand. Everyone

seemed to be in a hurry and a rage. The head cook, a fine, scarlet man with big moustachios, stood in the middle booming continuously, '*Ça marche deux oeufs brouillés! Ça marche un Chateaubriand aux pommes sautées!*' except when he broke off to curse at a *plongeur*. There were three counters, and the first time I went to the kitchen I took my tray unknowingly to the wrong one. The head cook walked up to me, twisted his moustaches, and looked me up and down. Then he beckoned to the breakfast cook and pointed at me.

'Do you see *that*? That is the type of *plongeur* they send us nowadays. Where do you come from, idiot? From Charenton, I suppose?' (There is a large lunatic asylum at Charenton.)

'From England,' I said.

(From *Down and Out in Paris and London* by George Orwell.)

Task 1

Now answer the following questions:

1 For how many hours did Orwell work during the day?
2 Which sentence tells you that he disliked getting meals?
3 Which words make the kitchen sound like a scene from Hell?
4 Why was the metal work covered with cloth?
5 What do you think a *plongeur* was?
6 Why was sand being used?
7 What are *oeufs brouillées* and *un Chateaubriand aux pommes sautées*?
8 Why did the head cook call the writer an idiot?
9 What would we call a *lunatic asylum* today?
10 Compare Orwell's description of a kitchen with the picture of the modern kitchen below. What are the main differences? Can you spot one thing that would be the same?

Task 2

Drawing from your own experience, write an account of a typical day working in a modern kitchen.

Assignment 4: The soft drinks survey

Task 1

Read the information on cola drinks taken from *Which?* magazine, and answer the following questions.

1. Coca-Cola was the invention of
   (a) Caleb D. Bradham.
   (b) Jacob's Pharmacy, Atlanta.
   (c) John S. Pemberton.
2. Pepsi-Cola was originally named
   (a) French Wine Cola.
   (b) Brad's Drink.
   (c) Eureka.
3. Most cola drinks are largely made up of
   (a) sugar.
   (b) flavouring.
   (c) fizzy water.
4. Soon after the year 1900, steps were taken to remove traces of
   (a) cocaine.
   (b) caffeine.
   (c) coca leaf.
5. Cola nuts contain
   (a) caffeine.
   (b) caramel.
   (c) cocaine.
6. During one year, the inhabitants of the U.K. each drank an estimated
   (a) 40 cans a day.
   (b) 40 cans a year.
   (c) 40 cans a week.

Task 2

Your employer, a 'soft drinks' manufacturer, wants to conduct a survey on soft drinks consumption in order to improve his sales. He asks you to design a questionnaire for random distribution as part of his market research, and to present your findings in the form of a short report. He wants the following questions answered:

- What other soft drinks do you buy besides cola drinks?
- Do you prefer cans or bottles?
- How many soft drinks do you consume during a week?
- Do you ever 'bulk buy' soft drinks?
- How much are you influenced by advertising and packaging?

COLA DRINKS

Does your family have a fierce brand loyalty to Coke or Pepsi? Or would a cheaper brand go down just as well? There's something for all age groups here: analysis of how much sugar and caffeine cola drinks contain; prices and taste tests; as well as some experiments younger members of the family may enjoy trying for themselves.

HOW IT ALL BEGAN..

In 1886, a new flavoured syrup went on sale at Jacob's Pharmacy, Atlanta, Georgia. The syrup was the invention of John S. Pemberton, a pharmacist whose other products included 'French Wine Cola – Ideal Nerve Tonic'.

Legend has it that fizzy water was accidentally added to the syrup instead of plain water, and so Coca-Cola's potential as a soft drink was discovered.

Pepsi-Cola appeared a few years later. The syrup was also the invention of a pharmacist, Caleb D. Bradham of North Carolina, and was first known as 'Brad's Drink'. In 1898 Bradham renamed it Pepsi-Cola.

Since those early days, many other cola drinks have come onto the market, and colas are sold all over the world.

WHERE COLA COMES FROM

Cola extract comes from the nut of the cola tree. The tree is native to West Africa but is grown in other parts of the world too. Cola extract has a bitter taste because the nuts contain caffeine.

WHERE THE 'COCA' COMES FROM

Flavour prepared from the coca leaf was used in the original Coca-Cola. Minute traces of cocaine may have been present in the early days, but soon after 1900 steps were taken to make sure that any such traces were removed from the flavour.

WHAT MOST COLA DRINKS ARE MADE OF

FIZZY WATER (water and carbon dioxide. About 90%, though, of course, it's not shown in proportion here).

SUGAR (about 5 to 11%) and/or **ARTIFICIAL SWEETENERS**.

CARAMEL (for colour).

FLAVOURING (including cola nut extract).

ACID (usually phosphoric).

EXTRA CAFFEINE

Preservatives and **other permitted additives** may be included.

The exact flavouring that goes into cola drinks is a closely guarded secret among manufacturers. You probably won't see cola extract mentioned on the label. That comes under the heading 'flavouring' and manufacturers don't have to be any more specific than that. All cola drinks should contain some cola extract. If they don't, they should be labelled 'cola flavour' drink. Other things used for flavouring cola drinks include citrus oils, vanilla, and spices like cinnamon and nutmeg.

EXPERIMENT

Leave a tarnished coin overnight in a glass of cola. By morning it will be bright again. That's because the acid in the drink dissolves the tarnish. Cola drinks aren't the only things that will do this – you can try it with other foods and drinks that contain acid, like lemonade and vinegar.

HOW MUCH COLA DO YOU DRINK?

Last year in the UK we drank an estimated 748 million litres of cola. That's an average of 40 cans per person. Do you drink more or less than other people? Some people drink a lot more. In Los Angeles, there's a clinic that treats 'colaholics' – people who drink up to 40 cans a day!

Assignment 5: The break-down

Mr. J. Bowen has arrived at your hotel *The Red Dragon*, Cheltenham, Gloucestershire, at 21.30 hours one Thursday. He is rather flustered because his car, a Renault, has broken down and he has to be in Stratford-upon-Avon the following afternoon. Using the information from an A.A. Handbook, do the following tasks.

CHELMSFORD 58,320 Essex Map 19TL70 EcWed MdTue/Fri/Sat cattle Fri *Cambridge*41 *Colchester*24 *Ipswich*41 *London*37 *Norwich*82 *Southend*20
★★★**South Lodge** 196 New London Rd ☎64564 17rm(16⇌9⁋) A16⇌ 45P B&B*(d)(e)*
★★**County** Rainsford Rd ☎66911 31rm(17⇌7⁋) A25rm(7⇌) B&B*(b)(c)*
¶¶*Signpost Mtrs* 112 Parkway ☎51201 ⏚ ✠ Aud VW
†¶¶*G S Last* 144 Moulsham St ☎61822 ☏65963 ✠ Tal
¶ ⏚*Oaklands S/Sta* Princes Rd ☎353720 ☏Basildon 412551
¶⏚*Oasis Autopoint* Springfield Rd ☎57047 ☏468317 R24hrs
CHELTENHAM 84,014* Glos Map17SO92 EcWed/Sat MdThu *Cirencester*16 *Evesham*16 *Gloucester*8 *London*96 *Oxford*41 *Tewkesbury*9
☆☆☆☆**Golden Valley** Gloucester Rd TS ☎32691 103⇌⁋ 300P
★★★★**Queen's** The Promenade THF ☎514724 77⇌ 32P B&B*(e)*
★★★**De La Bere** Southam (3m NE A46) ☎37771 *L* 22rm(20⇌2⁋) A11⇌ P B&B*(d)*
★★★**Carlton** Parabola Rd BW ☎514453 49rm(48⇌1⁋) 17P B&B*(d)*

Task 1

Find the telephone number of the most appropriate garage which could collect his car from outside your hotel.

★★★⏚⚘**Greenway** Shurdington ☎862352 CC RS 12⇌ ✗ 35P HBL nc7 B&B*(e)*
★★★**Lilley Brook** Cirencester Rd, Charlton Kings ☎25861 40rm(22⇌18⁋) 200P B&B*(d)*
★★★⏚**Savoy** Bayshill Rd ☎27788 56rm(26⇌3⁋) 21P B&B*(b)(c)*
★★★**Wyastone** Parabola Rd ☎22659 13rm(5⇌8⁋) ✗ 14P *B* B&B*(d)*
★★**George** St George's Rd ☎35751 46rm(7⇌⁋) 15P B&B*(d)*
★★⏚**Lansdown** Lansdown Rd ☎22700 13rm(8⇌5⁋) ✗ 30P nc8 B&B*(e)*
★★**Prestbury House** The Burgage, Prestbury (2m NE A46) ☎29533 10rm(5⇌) 40P B&B*(c)(d)*
★**Overton** St Georges Rd ☎23371 12rm(4⇌) 20P B&B*(c)(d)*
★**Royal Ascot** Western Rd ☎513640 9rm(1⇌) 30P B&B*(b)*
¶¶¶⏚*Bristol Street Mtrs* 83-93 Winchcombe St ☎27061 ☏510887 ✠ Frd
¶¶¶⏚*Lex Mead* Princess Elizabeth Way ☎20441 ☏20440 ✠
¶¶¶⏚*Mann Egerton* Imperial House, Montpellier Spa Rd ☎21651 ☏510886 *(c)* Ren
¶¶⏚*DG Ltd* Kingsditch Ln ☎28945 ⏚ ✠ Peu
¶¶⏚*Lyefield* 21-23 Lyefield Rd West, Charlton Kings ☎21131 ✠
¶¶⏚*Naunton Park* (New Victory Mechanics) Churchill Rd ☎26979 ☏20270
¶¶*Pihlens Mtrs* 60/66 Fairview Road ☎513880 ⏚ ✠ Dat

Abbreviations and Symbols

Garage entries

†	details not confirmed	⌘	motorcycle and/or scooter repairs undertaken
\	garage classification (see page 19)	♣	approved vehicle testing station at time of going to press; it is advisable to confirm by telephone
⋈	Free Breakdown Service classification; service normally available 24 hours every day, unless otherwise shown (see page 19)		
		mdnt	service until midnight
⊷	motorcycle specialist classification (see page 19)	R	repairs and servicing available *outside* normal working hours until time shown
⋈	Free Breakdown Service available Monday–Friday during normal working hours, unless otherwise stated	Vau etc	abbreviations for franchises held by garages (see page 92)

Garages
(fictional example)

Vehicle franchises The garage entries throughout the gazetteer include abbreviations of the makes of vehicle for which franchises are held.

Cars

AC	·AC	FSO	FSO	RR	Rolls Royce/Bentley	CZ	CZ
AM	Aston Martin	Hon	Honda	RT	Rover/Triumph	Duc	Ducati
AR	Alfa Romeo	Jep	Jeep	Sab	Saab	Gar	Garelli
ARO	ARO	Lad	Lada	Sko	Skoda	Gil	Gilera
Aud	Audi/NSU	Lnc	Lancia	Sub	Subaru	Hon	Honda
BL	Austin/Morris	Lot	Lotus	Suz	Suzuki	Jaw	Jawa
BMW	BMW	LR	Land Rover	Tal	Talbot (formerly Chrysler)	Kaw	Kawasaki
Bri	Bristol	Mas	Maserati			Lam	Lambretta
Bui	Buick	Maz	Mazda	Toy	Toyota	Mal	Malaguti
Che	Chevrolet	MB	Mercedes Benz	Vau	Vauxhall/Bedford	Mbc	Moto Becane
Cit	Citroen	MG	MG	Vlo	Volvo/Daf	Mgz	Moto Guzzi
Col	Colt	Mgn	Morgan	VW	Volkswagen	Mmr	Moto Morini
Dai	Daihatsu	Opl	Opel			MZ	MZ
Dat	Datsun	Peu	Peugeot	*Motorcycles*		Puch	Puch
DJ	Daimler/Jaguar	Por	Porsche	Bat	Batavas	Suz	Suzuki
Fer	Ferrari	Rar	Range Rover	Bet	Beta	Tri	Triumph
Fia	Fiat	Rel	Reliant	BMW	BMW	Ves	Vespa
Frd	Ford	Ren	Renault	BSA	BSA	Yam	Yamaha

Hotel entries

★	hotel classification (see page 25)	RS	restricted services operate for a period (see page 91)
★	hotel classification (see page 25)	U	unlicensed
☆	hotel classification (see page 25)	rm	number of bedrooms (see page 91)
⊕	approved hotel (see page 25)	⇨⏷	private bathroom and/or shower with own toilet (see page 91)
○	hotel likely to open during the currency of the *Handbook*	A	annexe (followed by number of rooms)
⚜	country-house hotel (see page 25)		
HBL	merit award (see page 25)	⚡	no dogs
❀	rosette award (see page 25)	P	parking on hotel premises (number of cars usually stated)
Ⓖ	mainly grill-type meals	P̸	no parking available on hotel premises
THF etc	abbreviations for hotel groups (see page 91)	nc	no children *eg* nc4 = no children under 4 years of age
C	closed for two months or more within a year (see page 91)		
CC	closed for less than two months at any one time (see page 91)		

Task 2

Fortunately the repairs needed for Mr. Bowen's car were done very quickly, and it was returned at the end of the following morning. In the meantime he has asked you if you could supply him with a route to Stratford-upon-Avon. He would prefer not to travel on a primary route, but nevertheless wants a fairly fast road.

Motorways with junction numbers and service areas

Motorway junction with restricted entry or exit.

Motorways under construction

Dual carriageways

Primary routes

Other A roads

B roads. Unclassified roads

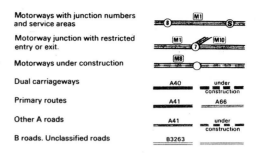

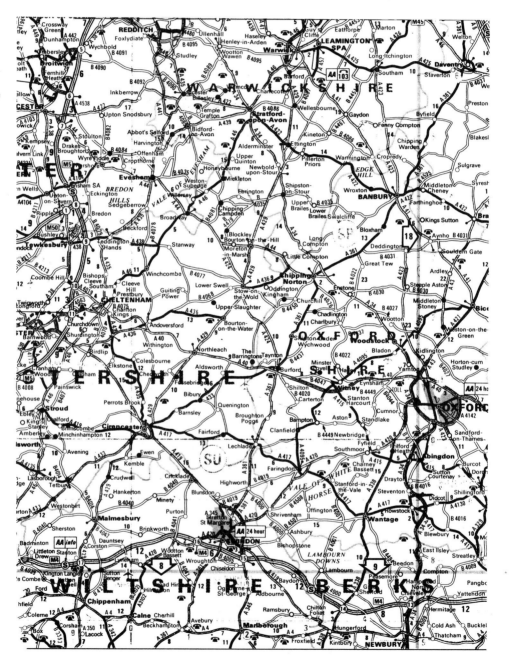

Task 3

Mr. Bowen wants to stay in a small hotel in Stratford-upon-Avon; but since he hates dogs, it must be one where dogs are not allowed. Find an appropriate hotel for him.

STOKE POGES 4,898 Bucks Map18SU98 EcWed Beaconsfield6 *London24* Slough3
⟋⟋*Stoke Poges Mtrs Co* Bells Hill ☎Farnham Common 2365 ⥁ ⚡ Ren
STOKESLEY 3,007 N Yorks Map41NZ50 EcWed MdFri/cattle Mon *London242 Middlesbrough10* Stockton-on-Tees10 Thirsk18
⟋⟋*Nesham Teesside* Meadowfield ☎710386 📞Gt Ayton 722741 ⚡ Frd
STONE 10,830 Staffs Map24SJ93 EcWed MdTue/cattle Thu *London149* Stafford8 Stoke-on-Trent8
★★★ **Brooms** ☎815531 11⇌⚐ ⚑ 40P B&B*(d)*
★★★B **Crown** High St ☎813535 13⇌ A16⚐ 154P B&B*(d)*
⟋⟋*Norton Bridge* (Robert Simcock & Sons) Norton Bridge ☎760281 R24hrs ⚡ BL
⟋⟋*W G Stone Motorist Centre* Crown St ☎815897 📞815033 ⥁ ⚡
STONE CROSS E Sussex Map10TQ60 Eastbourne4
STONEHOUSE 5,893 Glos Map17SO80 EcThu MdMon *Gloucester9* London111 Stroud3
⟋⟋*J A Gordon & Sons* Ebley Rd ☎2139 ⥁ ⚡ Tal
STONEYCOMBE Devon Map6SX86 *Exeter21 London189* Newton Abbot3 Torquay5
★★ **Bickley Mill Inn** ☎Kingskerswell 3201 10⇌ 80P B&B*(b)*
STONEYCROFT Merseyside Map32SJ39 **See Liverpool**
STONEY CROSS Hants Map8SU21 Cadnam3 *London87*
★★ **Compton Arms** NCC ☎Cadnam 2134 12⇌ 55P B&B*(c)*
STONY STRATFORD Bucks Map18SP74 **See Milton Keynes**
STORRINGTON 3,277 W Sussex Map10TQ01 **See also** Thakeham and West Chiltington EcWed *Brighton18* Guildford32 *London53* Petworth11 Worthing10
⟋⟋*Harwoods* The Square ☎3282 📞24hrs ⚡ BL
STOTFOLD 6,721 Beds Map19TL23 Baldock3 Bedford15 *London41* Luton14
⟋⟋*Stotfold Mtr Centre* 28 Astwick Rd ☎Hitchin 730222 R22.00 ⥁ ⚡
STOURBRIDGE 56,530 W Midlands Map26SO98 EcThu MdMon to Sat *Birmingham14* Kidderminster7 *London137* Wolverhampton10
★★ **Bell** Market St WDB ☎5641 21rm 100P B&B*(c)*
★★ **Talbot** High St WDB ☎4350 20rm(6⇌) 30P B&B*(d)*
⟋⟋*North Worcestershire Mtrs* 131-135 Hagley Rd, Oldswinford ☎3031 📞5881 ⚡ Vau AR
⟋⟋*Ring Road S/Sta* St Johns Rd ☎2266 📞75554 ⥁ ⚡
STOURPORT-ON-SEVERN 19,430 Heref & Worcs Map17SO87 EcWed *Birmingham22* Kidderminster4 *London146* Worcester12

★★★ **Mount Olympus** 35 Hartlebury Rd ☎77333 42rm(37⇌5⚐) ⚑ 150P B&B*(d)*
⟋⟋*Silver Seal Auto Centre* Worcester Rd ☎77262 ⥁ ⚡ Ren
⟋⟋*Lloyds* Bridge St ☎2053 📞3941
STOWMARKET 9,020 Suffolk Map20TM05 EcTue MdThu/Sat cattle Thu Bury St Edmunds14 *Cambridge43* Colchester28 Ipswich12 *London91* Norwich38
STOW-ON-THE-WOLD 1,737 Glos Map17SP12 EcWed Cheltenham18 Evesham16 *London84* Oxford29 Stratford-upon-Avon21
★★★ **Fosse Manor** ☎30354 17rm(4⇌6⚐) A6rm(2⇌) 50P B&B*(b)*
★★★ **Royalist** Digbeth St ☎30670 9rm(4⇌1⚐) A4rm(1⇌3⚐) 8P B&B*(b)(c)*
★★ **Stow Lodge** The Square ☎30485 C 20⇌⚐ ⚑ 30P nc5
★★ **Talbot** Market Sq CH ☎30631 32rm(8⇌) 10P B&B*(b)*
★★ **Unicorn** Sheep St CH ☎30257 20rm(17⇌3⚐) 50P B&B*(d)*
★ **King's Arms** Market Sq ☎30364 8rm 9P B&B*(b)*
⟋⟋*Parklands* Park St ☎31133 ⥁ ⚡ Dat Frd
⟋⟋*Stow S/Sta* Fosseway ☎31009 📞31458
STRADBROKE 880 Suffolk Map20TM27 EcMon/Thu Diss9 Ipswich28 *London112* Norwich28
⟋⟋*Stradbroke* Queen St ☎516 📞407 ⥁ ⚡
STRATFORD ST ANDREW 131 Suffolk Map20TM35 Aldeburgh8 *Colchester35* Ipswich18 *London96* Norwich38
⟋⟋*Murray* ☎Saxmundham 2516 ⥁ ⚡
STRATFORD-UPON-AVON 19,760 Warwicks Map17SP25 EcThu MdFri/cattle Tue Banbury20 *Birmingham24* Coventry19 *Gloucester39 London95* Oxford40
★★★★L **Hilton** Bridgefoot ☎67511 253⇌ 350P B&B*(e)(f)*
★★★★ **Shakespeare** Chapel St THF ☎294771 66⇌ 35P B&B*(e)*
★★★★ **Welcombe** Warwick Rd BT ☎295252 84⇌160P H B&B*(e)*
★★★ **Alveston Manor** Clopton Bridge THF ☎4581 116⇌ 200P B&B*(e)*
★★★ **Arden** Waterside INT ☎294949 63rm(41⇌2⚐) 40P B&B*(b)*
★★★B **Falcon** Chapel St GM ☎5777 73⇌⚐ 100P B&B*(d)*
★★★ **Grosvenor House** Warwick Rd BW ☎69213 57rm(23⇌30⚐) ⚑ 50P B&B*(c)(d)*
★★★ **Swan's Nest** Bridgefoot THF ☎66761 70rm(52⇌8⚐) 100P B&B*(d)*
★★★ **White Swan** Rother St THF ☎297022 56rm(22⇌2⚐) ₽ B&B*(d)*
★★ ⚞**Haytor** Avenue Rd ☎297799 RS 18rm(13⇌⚐) ⚑ 20P
⟋⟋*Heron* (Saville Mtrs) Birmingham Rd ☎67555 📞69968 ⚡

17

Guide-lines on preparing a short talk

1. When you know the subject of your talk, write down all the ideas, words and phrases that are relevant to the subject.
2. You will now have a very random and disjointed list. Arrange it into logical order.
3. The next step is to sort the order into sections. What are you going to talk about first? How are you going to develop the talk? How are you going to finish?
4. Your talk is beginning to take shape! Now, plan the introduction. This is very important. You must make your listeners *want* to listen to you and to feel that what you have to say is going to be worth while and enjoyable. The mood is usually established in the introduction, so plan it carefully.
5. Write the whole talk out in full, read it out loud and make any alterations that you feel

may be necessary. The actual writing out will help to commit the content and the shape of the talk to your memory. Now, with equal care, plan the conclusion which should bring your talk to a definite and memorable end. Avoid phrases like 'and that's about it, really!'

6 Consider the visual aids you will need for your talk: posters, pictures, photographs, diagrams, etc. Make sure that they are big enough to be seen and that the diagrams are bold. (You can easily check your visual aids in advance.) If you are using small photographs, mount them onto card and show them at the appropriate time during your talk, or invite your audience to come and look at them afterwards. Never pass photos or objects round while you are actually talking, otherwise your audience will stop listening.

7 If you are using any technical equipment, make quite sure that you know how it works! Prepare slides, etc. well in advance so that you show them the right way round and in the correct order.

8 Now transfer your talk, in heading form only, onto small pieces of card called 'cue cards'. These are to remind you of what you are going to say. Avoid trying to memorise the whole thing word for word, but do practise in front of friends if possible. If not, use a mirror. Preparation is very important indeed.

9 Stand comfortably, relax and look at your audience.

10 Talk *to* your audience, never *at* them. Smile at them – you have something to say which they will enjoy – share it with them. If you have prepared well, you will be confident. If you are confident, you will be enthusiastic, and if you have the combination of confidence and enthusiasm, the chances are that you will be very good indeed!

Guide-lines on preparing a demonstration

1 Make a list of all the pieces of equipment you will need, and make quite sure that they are ready for use i.e. clean, and in working order.
2 Lay your equipment out in an orderly way so that you don't have to search (or scrabble around in a bag) for a particular utensil.
3 Plan the introduction to your demonstration. (See note 4 of *Preparing a short talk*).
4 Once you start the demonstration guide your listeners through it step by step, avoiding long silent gaps as you demonstrate a particular skill. It's very important to rehearse your demonstration so that you know when the gaps are likely to occur and can prepare accordingly.
5 Don't panic if something goes wrong; audiences are amazingly sympathetic, especially if you take them into your confidence and explain how and why the mistake occurred and how you would correct it.
6 Plan the conclusion of your demonstration and remember that people love to sample food and drink, and to see clearly the end product whatever it might be.
7 Note **10** of *Preparing a short talk* applies equally to giving a demonstration – read it carefully.

Assignment 6: The careers evening

You have been approached by the Head of your former school to attend a 'Careers evening' and talk to a number of pupils who have expressed interest in a career within the catering industry. You have been allocated a room in the Domestic Science and Home Economics block for the purpose.

Task 1

Choose one of the following:
1 Prepare and give a demonstration of a skill learned during your training, to interest a group of school leavers.
2 Prepare and give a talk to a group of school leavers on *A Career In Catering*.

Note: Give yourself a time limit of 3 to 5 minutes for both of these.

Assignment 7: Trouble in the kitchen

The situation

You have recently started work as a commis chef in a restaurant which enjoys a high local reputation. The kitchen staff consists of a *head chef*, (chef de cuisine), *second chef*, (sous chef), a *patissier* who works in a separate kitchen alongside the main one, and *three commis chefs*, one of whom assists the patissier in the making of all the puddings, sweets and pastries. The restaurant staff consists of *head waiter, wine waiter, three waiters and three waitresses*.

The characters

The head waiter has worked in a well known Paris restaurant before coming to this country and although he was in no position of great authority, he frequently refers to his days in Paris, implying that the restaurant there was very superior to his present place of work. This infuriates the head chef who, although she cannot claim to have worked in Paris, is not convinced that this necessarily means very much, and she is quite confident that the standard of cooking and the quality of the dishes produced in her kitchen can compete with any restaurant anywhere. (This may not be strictly true, but nevertheless her standards are very high indeed.) The second chef is a quiet, good-natured man who was given his position because of his considerable ability rather than for his qualities of leadership. Both he and the head chef dislike the head waiter intensely. The patissier keeps very much to himself. His gateaux are famous in the area and he likes to be left alone to run his kitchen in his own way with a young commis as his assistant. The commis works well with the patissier who is an excellent teacher, and feels she is really learning a great deal. The other two commis have come straight from the catering department of the local College of Further Education. One is rather quiet and conscientious and tries hard to please; the other one tends to be careless although is capable of extremely competent work. Unfortunately, she also rather enjoys a 'spot of trouble'.

Task 1

In a group adopt the roles of the main characters. (Note that some of them – the waiters and waitresses, possibly the customers in the restaurant and the wine waiter – I have deliberately left for you to create.) Devise a 'spot of trouble' which occurs one day when the head chef is away ill.

Task 2

As an individual write to the head chef during her illness to tell her about the trouble, and how it was resolved.

Assignment 8: The caterer's kitchen knives

You have seen the following advertisement in a catering magazine:

> ## AMAZING OFFER!
>
> Set of cook's kitchen knives — superb quality only £15.00 (usually £25.00) complete in canvas hold-all.
>
> Money back if not fully satisfied.
>
> *Send now to:*
> **Caterers' Crafts Ltd.,
> 9, High Street,
> Bolton, Lancs. BN1 5BL.**

This appears to be a very good offer and you decide to send away for a set of these knives.

Task 1

Write the appropriate letter to Caterers' Crafts Ltd., ordering the knives.

Task 2

The knives arrive and at first you are very pleased with them. After one week, however, the handle breaks on one of them and you decide to send them back. Write a letter of complaint to Caterers' Crafts Ltd. asking either a) for a replacement set of knives or b) for your money back.

Task 3

Imagine that you are the Sales Manager of Caterers' Crafts Ltd., and reply to this dissatisfied customer.

Guide-lines on graphics

Apart from the quality of its food and service, the success of any catering outlet, whether it be a fish and chip shop or a five star hotel, depends on the way it presents itself to the general public. Hand-in-hand with the design of furniture, floor-covering, table-linen, cutlery and crockery, go the design of brochures, tariffs, menus, place-names and notices. This graphically designed material is often the initial selling point of any establishment. Remember — you read the menu first before you order the food! The following points are important:

1. Anything that is graphically represented must be easily seen and understood, so it must be eye-catching, readable and clear.
2. Beware of being too fussy. It is fashionable nowadays to present large hand-written menus that are often extremely difficult to read. The best designs are often the simplest. The following equipment and techniques will help you achieve a degree of professionalism.

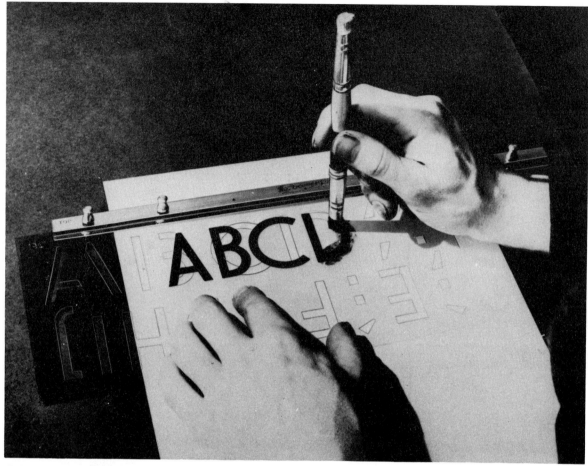

Dry brush stencilling

This is a very effective and simple method of lettering which has the advantage of being relatively cheap. It involves using a special brush which is moistened very slightly on a damp felt pad and then worked into colour until it is nearly dry. It is then rubbed over the stencil.

Pen stencilling

This is a sophisticated technique involving the use of special pens, templates and stencils. The equipment is expensive, but would be a worthwhile investment if it is going to be used extensively.

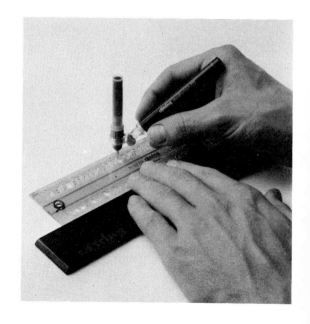

Dry transfer printing

This is a very popular form of printing using letters and symbols which are printed on to paper by simply rubbing them off with the rounded head of a pen or pencil. The result is extremely professional, and a tremendous variety of letters and symbols are available.

All these aids should be very familiar to art departments or graphics units of schools or colleges. However, information and equipment for *dry brush stencilling* is available from The Econasign Co. Ltd., 31, Britton Street, London. EC1M 5NQ. *Pen stencilling* equipment is produced by firms such as Rotring and Staedtler and may be obtained from any good stationer or art supplier. *Dry transfer* sheets produced by firms such as Letraset, similarly may be obtained from any good stationer or art supplier.

Assignment 9: The advertising campaign

One of the following has recently opened in your area:
- a fish and chip shop
- a small hotel
- a café (selling home-made cakes)

Choose which one you prefer and run an advertising campaign for it consisting of:

Task 1

Designing a full page advertisement for the local paper or magazine.

Task 2

Designing a car or bus sticker.

Task 3

Writing an advertising script for your local radio.

Task 4

Planning a special 'promotion' occasion.

Assignment 10: The sailing club dinner

The Arrow Sailing Club are holding their annual dinner at your hotel.

Task 1

Design a menu appropriate for the occasion.

Task 2

Design a place name.

Task 3

Write a letter to James Bros., Wholesale Fish Suppliers, The Wharf, Brixham, Devon, confirming a telephone order for 40 lobsters. (Remember the date and time!)

Task 4

Make a list of the people who would be involved with the dinner and explain how you would collaborate with them.

Assignment 11: Confirmation of booking

You have received an enquiry from a Mrs. J. A. Long, Acorn Cottage, Lower Wanswell, Herts, who wishes to stay at your hotel (*The Crown Hotel*, Dover, Kent), for the weekend of May 12th, 13th, 14th, with her husband. She wonders if you can also accommodate their dog.

Task 1

Write an appropriate letter making a provisional booking for Mrs. Long.
Note: The hotel does not allow pets.

Task 2

Design the hotel brochure you would send with the letter.

Assignment 12: The Falcon sales conference

The Falcon Car Company Ltd. has arranged to hold a Sales Conference for its top 50 sales representatives at your hotel from Friday evening, 25th March, until Sunday, 27th March, finishing with sherry at 12.00 noon. The number of people wanting accommodation will total 56 including the sales director and managing director of the company.

Task 1

Write a letter to the Sales Director, Falcon Car Company Ltd., 7, Earlsdon Court Road, Coventry, confirming the accommodation arrangements for the conference and enclosing a tentative programme for approval.
Note:
1 There will be a reception followed by dinner on the Friday evening after which the Sales Director will give a short speech.
2 There will be two conference sessions on Saturday with time allowed for meals and coffee and tea.
3 On Sunday there will be one closing session.

Task 2

Design the programme.

Task 3

Design the notices giving directions which will be put up in the foyer.

Task 4

Make a list of the other facilities which will be provided by the hotel, e.g. stationery.

Task 5

Write a memo to the hotel manager reminding him or her of the conference one week before it is due to take place.

Task 6

Provide the dinner menu for the Friday evening.

Assignment 13: A tourist package

The English Tourist Board is a statutory body created by The Development of Tourism Act, 1969 to develop and market England's tourism. The main objectives are to provide a welcome for people visiting England, to encourage people living in England to take their holidays there, and to encourage the provision and improvement of tourist amenities and facilities in England.

From a hand-out produced by
The English Tourist Board,
4 Grosvenor Gardens,
London. SWIW OOU.

Task 1

Look at the publicity advertisement for Bristol. Now using postcards, photos and possibly other materials, design a similar advertisement for your own area.

Task 2

Make a list of notices displayed in hotels which should be written in languages other than English. Which other languages should be used?

Shipshape and Bristol fashion

S.S. Great Britain
Isambard Kingdom Brunel's famous ship - the first iron ocean-going propeller-driven liner launched in 1843. Now being restored in her original dock.

Lochiel
Once a ferry between the Scottish Islands, now converted to a high class restaurant and pub.

Embassy Grand Prix
Speeds of over 100m.p.h. are reached on the most demanding and spectacular powerboat racing circuit in the world.

World Wine Fair
The major international event in the wine trade calendar held in the middle of the City's colourful dockside area with more than 30 countries exhibiting thousands of different wines.

Bristol Industrial Museum
Brings Bristol's maritime, aviation and industrial history to life. The National Lifeboat Museum is being established next door.

Boat Trips
Several companies operate boat trips round the harbour and on the River Avon. There are also occasional steamer trips into the Bristol Channel.

Bristol – A proud part of England's Maritime Heritage.

Further information from Director of Entertainment and Publicity, Colston House, Colston Street, Bristol BS1 5AQ. Tel: (0272) 26031.

Task 3

Look at the charts concerning visitors to the U.K. and answer the following questions:

1. Most visitors who came to the U.K. in the months of January to June 1984 came from
   (a) North America.
   (b) Other areas.
   (c) Western Europe.
2. There was a greater percentage change in visitors from
   (a) Western Europe.
   (b) North America.
   (c) Other areas.
3. The percentage change was greatest in
   (a) January.
   (b) March.
   (c) February.
4. Which year attracted the greatest number of overseas visitors?
   (a) 1978.
   (b) 1979.
   (c) 1983.
5. Which year was the best for hoteliers?
   (a) 1984.
   (b) 1975.
   (c) 1976.
6. What percentage of rooms in English hotels was occupied in June 1984?
   (a) 66%.
   (b) 72%.
   (c) 53%.

INCOMING TOURISTS

Overseas visitors to UK: % change 1984 over 1983, globally and by regions of origin

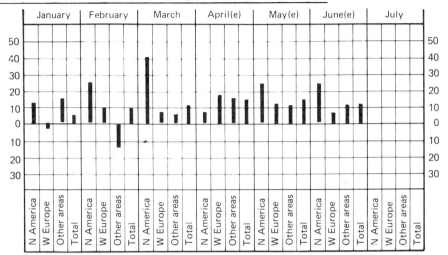

(estimates: percentage changes based on returns to nearest 10,000. Source: Department of Trade and Industry.)

Overseas visitors to UK

% Running totals January to June ('000)				Annual totals ('000)					
	1984	1983	% change	1983	1982	1981	1980	1979	1978
North America	1,296	1,075	+20.6	2,836	2,136	2,105	2,082	2,196	2,475
W. Europe	3,455	3,166	+9.1	7,199	7,082	7,055	7,910	7,873	7,865
Other areas	1,076	992	+8.5	2,464	2,418	2,291	2,429	2,417	2,306
Total	5,827	5,233	+11.4	12,499	11,636	11,452	12,421	12,486	12,646

OCCUPANCY

Average June occupancy of English hotels 1975–1984

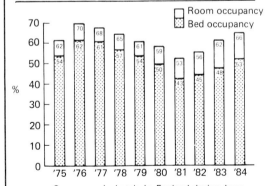

1984 monthly average occupancy

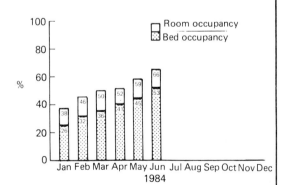

Occupancy in hotels in England during June 1984 improved for third consecutive year, taking room occupancy to average of 66 per cent and bed occupancy to 53 per cent. Bed occupancy in London averaged 72 per cent — improvement of 19 per cent points in two years.

Heart of England was only other region to show dramatic improvement on previous two years, and, like London, this region reported particularly high proportion of overseas arrivals. Conference hotels across the country accommodating 200-plus delegates returned occupancy average of 74 per cent in June 1984 (1983: 63 per cent; 1982: 58 per cent).

Source: English Hotel Occupancy Survey, English Tourist Board.

Assignment 14: *Love thy neighbour*

You have recently been made manager of a small café specialising in sea-food dishes in a popular seaside resort. Next door is a boutique selling clothes and local craft products. One day, during a particularly hot spell, you receive a strongly worded letter of complaint from the owner of the boutique next door claiming that the smell from your dustbins is affecting trade, and that the matter will be reported to the Environmental Health Officer.

You realise that the smell has resulted from the remains of shell-fish in dustbins that, on this particular occasion, have not been emptied on time because of the summer bank holiday, and the weather is very hot. However, you clearly must do something.

Task 1

Write the letter from the owner of the boutique.

Task 2

How do you think the Environmental Health Officer would deal with this situation?

Task 3

Write an appropriate letter back to your next door neighbour, bearing in mind that a friendly relationship between you is very important.

Task 4

Read the information on p. 37 from the consumer magazine *Which?* and find out the local addresses and telephone numbers of
1 the Citizens Advice Bureau.
2 the Consumer Advice Centre.
3 the Trading Standards Department.
4 the Environmental Health Department.
Find out how they operate. You never know, you may need these services one day.

Complaining about neighbours

LOVE THY NEIGHBOUR – IF ONLY YOU ALWAYS COULD

Neighbours are not just the people who live next door, but can be anyone who directly affects you by what they do. Most of your legal rights are based on the customs and practices which have grown up over the years. The rule is to try to live and let live.

Some problems are covered by the Control of Pollution Act 1974, and the Clean Air Act 1956. There may be additional local by-laws.

Your local environmental health officer may be prepared to take action through the criminal courts, if necessary. If not, you may have to take your own action in the civil courts. You usually have the options of trying to get an injunction to stop your neighbour doing whatever is annoying you, and suing for damages for injury or inconvenience caused. This can be costly, unless you're claiming less than £500, when you can go to a 'small claims' court and represent yourself.

BONFIRES

Mrs Cook's problem was a bonfire lit by neighbours to burn a felled tree. It burnt for five days, coating everything in the area with fine white particles. A phone call to the local council revealed that the area was not a smokeless zone, nor were there any by-laws relevant to her case. But the complaint was passed to the environmental health department and within two hours the bonfire was out.

The lesson for you
Even where there are no relevant local by-laws, a word from an official may help.

NOISY ANIMALS

Mr Poulter lives on a housing estate. His neighbours keep chickens and early every morning he's woken by cock's crow.

While cockerels in the country may have something to crow about, you've probably got grounds for complaint if they do it in the middle of the town.

Mr Poulter hasn't taken his case any further yet. But someone else we heard from, who was bothered by a dog howling in an upstairs flat, found that the matter was dealt with efficiently by the local environmental health officer.

The lesson for you
If a direct approach to your neighbour doesn't work, try contacting the environmental health officer.

SPRAYED WITH PAINT

Mr Smith lives next door to a paint spray plant. One day he noticed spots of grey paint on his car, his wife's car and the exterior paintwork of his house. The company agreed to clean both cars, but not the paintwork. The spots wouldn't come off and the cost of repainting the house was about £500. So, Mr Smith took action and eventually went to court.

He contacted his other neighbours who were prepared to support his claim. And he contacted the environmental health department of his borough council for help. At the hearing, Mr Smith won and was awarded £150 plus £32 costs. He thought this was reasonable, and the company was given two weeks to pay. They actually paid up after a month, just as Mr Smith was considering further action, and fifteen months after the problem first started.

The lesson for you
Taking the right action can pay off; it's also worth seeing if anyone else is affected, too.

OVERHANGING TREES

A lime tree overhanging Mr H's garden released sap over his front steps. He decided to take action after his child slipped and fell. Mr H chopped off the offending branches and stacked them in his neighbour's garden.

Two years on, Mr H noticed that the roots of the tree were now growing through his wall, making it lean at a dangerous angle. And the tree was still shading his garden and making the steps a hazard. He wrote to see if his local council could get his neighbour to do something.

The council agreed that the tree was causing damage and was a nuisance, but said they could not intervene unless the tree or wall became dangerous.

The lesson for you
Branches overhanging your garden belong to your neighbour. Contrary to popular belief, the same is true of windfall fruit that drops on your property.

The problem also shows that you may have to take your own action. If the worse came to the worst, Mr H could go to the courts and seek an injunction against his neighbour to keep the growth of the tree within the boundary, and to claim compensation.

WEEDS

If you're troubled by invading weeds, there's not a lot you can do unless you're covered by the Weeds Act, which exists to keep agricultural land free from certain named weeds.

Where to get help

Citizens Advice Bureaux (CABx) May be able to help on almost anything.

Consumer Advice Centres Free advice on consumer problems. If there isn't one in your area, try trading standards department.

Trading Standards Departments Mainly there to enforce the laws which control traders.

Environmental Health Department Enforce the various Acts and Regulations which apply to food, health and safety, pollution, noise, housing.

Legal Advice/Law Centres Free legal advice from qualified lawyers on most areas of the law.

Local Councillors Your local councillor is a good person to contact for help in disputes with the local authority. You have to go through a councillor if you're complaining to the Local Government Ombudsman.

MPs Can be an effective route for complaining about issues of more general importance and personal difficulties with authorities.

Solicitors' Advice At a price unless you have a fixed fee interview, £5 for 30 minutes, or you qualify for legal aid.

Which? Personal Service Costs £20 a year. A team of lawyers is available to advise on consumer problems. Where necessary, Which? Personal Service can help you take your case to court or intervene on your behalf with the supplier of faulty goods. For further details of how to join write to Dept BJJ, 14 Buckingham Street, London WC2N 6DS.

Publications
Which? way to complain, by Ian Cooper (Consumers' Association), £4.95. Fuller guide to making effective complaints, with relevant addresses.

A Handbook of Consumer Law, Federation of Consumer Groups, (Consumers' Association and Hodder and Stoughton), £3.95.

Small Claims in the County Court, by Michael Birks (Form EX50) available free from county courts and CABx.

OFT publications
The Office of Fair Trading produces a string of useful leaflets (mostly free) available from advice centres.

Which? reports
The Telephone Service, February 1979 (new report planned for next month). *Dentists*, November 1979. *Dealing with the Gasman*, May 1982. *Getting Legal Advice*, October 1982. *Legal Expenses Insurance*, *Money Which?*, December 1982. *Consumer Law: What are your rights?*, May 1983. *GPs*, June 1983. *Problems with your accountant, solicitor, architect or surveyor*, July 1983. *Insolvency*, August 1983. *TV – the Viewers' View*, November 1983.

Assignment 15: The hotel flowers

You are a trainee receptionist at the *Imperial Hotel*, Llandudno, Gwynedd, and the head receptionist has handed you a brochure which was received in the post from a firm specialising in commercial floral arrangements, (for shops, offices, hotels, etc.). The receptionist thinks it would be a good idea to investigate the possibility of having the flowers for the hotel foyer and dining room arranged by this particular firm: Floral Arts Ltd., 5, Chester Road, Wrexham, Clwyd.

Task 1

Write an appropriate letter to the firm inviting their representative to call.
As a result of the visit you decide to employ Floral Arts for a trial period of six months.

Task 2

Write the appropriate letter to Ms. A. Smith of Floral Arts confirming this arrangement.

Task 3

Write a memo to the hotel manager giving the details of the decision.

Task 4

At the end of six months you decide that the services of Floral Arts Ltd. have declined and that you would rather arrange the flowers yourselves. Paying particular attention to the tone, write the appropriate letter.

Guide-lines on writing a short report

Don't be put off by the word 'report'. Remember that, quite simply, a report is written *to* someone, *by* someone, *about* something. It should be as simple and straightforward as possible. Because it is the result of careful investigation, it must be arranged in some sort of logical order. It must be readable and therefore it should be attractively presented. Remember that if you find it boring to write, it will probably be boring to read!

The structure

A report should be arranged as follows:
1 Title page – title of subject and name of author.
2 Aim – a brief, clear statement of the purpose of the report.
3 Contents – a list of the sections of the report.
4 The main body of the report arranged in logical order.
5 Conclusion.
6 Recommendations.
7 Bibliography – a list of books and publications used.
8 Appendices – any tables, graphs and diagrams which are 'extra' and not easily inserted into the main text should be included in this section at the end of the report.

Assignment 16: A consumer report

You are cleaning the residents' lounge of the small family hotel where you work, when the very ancient vacuum cleaner you are using breaks down. The manager asks you to buy a new one, but you must do some research first.

Task 1

Either by writing to particular manufacturers, or by direct approach to retailers, find out what cleaners are available, and select the one most suitable for the needs of your particular hotel.

Task 2

Present your findings to your manager in the form of a short report justifying your choice.

Assignment 17: An accident report

You, and another member of staff are running the bar one evening when some beer is accidentally spilt on the polished tiled floor behind the bar. Your colleague slips on the beer, cuts her leg which subsequently needs four stitches, and badly sprains her ankle.

Task 1

Write out a full report of the accident for the insurance company giving:
(a) Date
(b) Time
(c) Place
(d) Your own account of what happened, using a sketch plan if necessary.
(e) Details of first-aid given.
(f) Names and addresses of two witnesses who can be contacted if necessary.
(g) Your opinion of the causes of the accident and remedies.

Task 2

You are asked by your manager to do some research into floor coverings and surfaces for use in a bar. This necessitates looking at trade literature and possibly writing away for information. You should present your research in the form of a short report.

Assignment 18: A hotel in Brittany

The following is a page from Arthur Eperon's Guide to the best of French hotels *The French Selection*. Please read it carefully.

For many years this has been one of my very favourite seaside hotels anywhere, because of its position, its tasteful furnishings, the charm of its building, its excellent cooking of fresh local food, and above all for its atmosphere and the welcome from Gérard and Danielle Jouanny.

A granite manor house, hidden from the road by its trees, drive and lawns, it stands high above the bay and yacht harbour with great views from its terrace to the little islands, and in good weather to the north coast of Finistère. The grounds drop sharply below the terrace through trees and shrubs with a private path down to the beaches. You get the same superb views from the sun lounge-style dining room. The hotel is very pleasantly furnished, mostly with old pieces. The cosy lounge has a pleasant tapestry, there is a real bar outside the dining room. All bedrooms are different and are tastefully decorated. Most have big windows and all have bathrooms. Some have balconies looking to sea and two have small sitting rooms.

Although this is a Relais du Silence, the family are very helpful in looking after children and provide a children's meal; but I don't think Trébeurden is the best place for toddlers because of its steep hills – unless you want to drive every time you go to the beach. It's a pleasant little old-style resort, excellent for children about seven and over. It has several crescents of fine sand divided by rocks. Two are divided by a rocky peninsula, Le Castel, joined to the mainland by a strip of sand. Westerly winds make it a splendid place for sailing and in summer the bay is sprinkled with sails of many colours. You can take a boat to the isle of Millau, from which you can see a great stretch of the coast. There is a golf course.

Task 1

Now answer the following questions:
1 What is the chief reason given for this hotel being one of the author's favourites?
2 What view can be seen from the terrace of the hotel?
3 Which words tell you that the hotel stands on high ground?
4 What kind of furniture would you find in the hotel?
5 How many bedrooms have sitting-rooms adjoining?
6 Using the resources of your college library try and find out what a Relais du Silence is.
7 Which words tell you that children are catered for in the hotel?
8 Why is the hotel unsuitable for toddlers?
9 What is Le Castel?
10 What outdoor activities can be enjoyed from the resort?

Task 2

Write an entry for a tourist guide about a restaurant, inn or hotel known to you.

Assignment 19: A weekend in the country

You have recently started work at a country hotel and the manager has asked you to prepare an advertisement as part of a promotional campaign.

Task 1

Design the advertisement for either a newspaper or magazine.

Task 2

Design a brochure and tariff to send out to prospective guests.

Task 3

Re-design the existing rather dull menu.

Task 4

As a result of your advertisement you receive a letter from a Ms. J. Immington of English Country Tours Ltd., inquiring if you can accommodate 30 wives of businessmen attending a conference in London, for a weekend (the date will be your choice). They will arrive from London at 21.00 hours on Friday and depart after breakfast on Monday. Ms. Immington has also asked if you could suggest a weekend itinerary for the 30 ladies. You are obviously very keen to win the custom of English Country Tours Ltd. and so you write an appropriate letter back to Ms. Immington enclosing:

(a) the hotel brochure and tariff,
(b) a suggested weekend's itinerary,
(c) a sample menu.

Assignment 20: The seminar

Ms J. Immington of English Country Tours Ltd., is holding a seminar at your hotel in order to promote your area as a Tourist Centre. The seminar will last for one day only and the hotel will provide morning coffee, pre-lunch sherry, lunch and afternoon tea.

Task 1

Design suitable notices for the hotel foyer telling guests where the seminar is being held and giving directions.

Task 2

Make a list of all the facilities the hotel should provide for such a seminar.

Task 3

Re-design the lunch menu to suit the occasion.

Task 4

Write a letter to Ms Immington confirming her booking of your hotel for the seminar and telling her what contribution the hotel is making towards it in the way of facilities, etc.

Task 5

Write a memo to the hotel manager a week before the event, reminding him of it.

Task 6

Prepare a small display for use in the hotel foyer advertising the attractions of your area.

Assignment 21: Easter in Paris

It is a particularly dreary and grey day at the end of January. The delights of Christmas have now disappeared into the past, but the Spring is still far away into the future. One cold, wet lunch hour you happen to glance into a travel agent's window and see a large poster advertising an Easter Break in Paris. It looks very attractive and you decide to investigate further. You enter the travel agent's and pick up the appropriate brochure.

Task 1

Study the information given and answer the following questions:
1. If you decided to take your own car with you, how long would it take you (approximately) to cross the Channel by hovercraft and drive to Paris?
2. You decide to spend four nights in Paris, staying in an hotel described in the brochure

as 'budget tourist grade' and your rooms (there are four of you), have a bathroom or shower adjoining. How much would the basic cost be for each of you?
3 If you went by train and took an overnight boat when would you:
(a) leave London?
(b) arrive in Paris?
(c) How long would the journey take from London to Paris?
4 If you decided to travel by air from Gatwick or Heathrow, at which Paris airport would you land?
5 Name another airport in Paris.

6 What is the tour number of the cheapest Easter break?
What is the tour number of the most expensive?

Task 2

Using the facilities of your college library or other sources, write a brief paragraph on each of the following:
1 Charles de Gaulle.
2 Les Bateaux Mouches.
3 Le Centre Pompidou.
4 The Eiffel Tower.
5 The 'Left Bank'.

Easter and all the Bank Holidays for 1985

Tour No	Travel Route	EASTER depart Thursday or Friday 4 NIGHTS PH1 PH2 PH3 PH4	EASTER depart Friday 3 NIGHTS PH1 PH2 PH3 PH4	BANK HOLIDAYS (May, Spring & August) depart Friday 3 NIGHTS PH1 PH2 PH3 PH4	BANK HOLIDAYS depart Saturday 2 NIGHTS PH1 PH2 PH3 PH4
PCE	**Paris Coach Express**	not available	63 72 79 –	61 70 77 –	not available
BR1	**Rail** - ship or hovercraft - day services	74 85 94 110	66 76 83 95	64 74 81 93	59 66 71 78
EXO	**Rail** - ship overnight service (3 nights hotel)	68 79 89 99	not available	not available	not available
PAE	**Paris Air Express**	not available	79 88 94 104	77 86 92 102	not available
BCA	**British Caledonian** (Gatwick)	106 118 125 142	100 109 114 127	98 107 112 125	92 98 102 110
BAF	**British Airways or Air France** (Heathrow)	111 123 130 147	105 114 119 132	103 112 117 130	97 103 107 115
TYC	**Take your own car** (each of 4 persons)	64 76 84 100	58 67 73 85	56 65 71 83	50 56 60 68

Hotel grade explanation: PH1: Budget tourist grade without bath/shower
(For details see pages 6/7) PH2: Budget tourist grade with bath/shower
  PH3: Tourist grade with bath/shower
  PH4: Higher grade with bath/shower

Single rooms: PH1 £4; PH2 £6; PH3 £8; PH4 £10 per night.
Twin rooms: £1 per person per night.

Take your own car is based on 4 persons per car. Supplements for only 3 in a car £6 each; only 2 in a car £16 each. August Bank Holiday supplement £22 per car. Prices are in £'s per person in double room (large bed) according to Hotel grade and travel route selected.

ITINERARY: Rail/ship overnight services (Easter only)

Depart London late Thursday evening (approx. 9p.m.) via Newhaven - Dieppe. Ship crossing approx. 4 hrs. Arrive Paris, St. Lazare approx. 7a.m. Friday morning. Coach transfer to hotel and breakfast. Three nights hotel. Return similar timings on Monday evening arriving London Tuesday morning.

ITINERARY: Scheduled Air from Gatwick or Heathrow

Depart Gatwick or Heathrow mid-morning or afternoon to Paris, Charles de Gaulle airport (1 hr). Make your own way from the airport by bus or train (27Frs.) to central Paris and taxi or Metro to your hotel. Return on afternoon or evening services.

ITINERARY: Take your own car

Depart Dover at a time to suit your driving time from your hometown. Please state clearly on the Booking Form your earliest preferred departure from and latest return to Dover. Cross channel by hovercraft to Calais or Boulogne and approx. 3½ hrs. drive into Paris. Parking cannot be guaranteed.

ITINERARY: Paris-Air Express

Depart Gatwick, early afternoon Friday by Dan-Air 1-11 Jet for Beauvais (40 mins.) and express coach direct from the airport to your hotel arriving early evening. Three nights hotel. Early afternoon departure on Monday from central Paris terminal arriving Gatwick late afternoon.

ITINERARY: Daytime rail/ship or hovercraft

Depart London mid-morning or midday by train (with reserved seats) for Dover. Ship crossing approx. 1½ hours, hovercraft approx. 35 mins. Arrive Paris, Gare du Nord late afternoon or early evening and transfer by coach to your hotel. Return similar timings arriving London afternoon or evening.

ITINERARY: Paris Coach Express

Depart London (King's Cross Coach Terminal) at 9.30a.m. Friday for Dover. Ship crossing approx. 1½ hours. Arrive Paris hotel approx. 10p.m. Three nights hotel. Saturday morning tour of Paris and afternoon visit to Versailles also included. Return London late Monday, approx. 10p.m.

Assignment 22: A visit to London

You are working in a small restaurant which has recently opened in Herefordshire. The restaurant is called *The Country Kitchen*, and the manager is trying to create a decor to suit the name. You are intending to visit London for the January Sales, and the manager asks you to visit a shop near Covent Garden which specialises in well-designed kitchen ware, in order to buy some earthenware terrines for *The Country Kitchen*. You decide to travel to London on the following Friday from your nearest station, Ledbury.

Task 1

Study the time-table and answer the following questions:
1. Which train would you catch in order to reach London at about 10.00 am?
2. At which London station will you arrive?
3. Will you be able to have a cooked breakfast on the train?
4. Your friend wants to visit a relation in Pershore the same morning. Which is the earliest train she can catch from Ledbury?
5. What does the letter R mean beside the name of Hereford and Worcester Shrub Hill?
6. What is the name of the service identified by the initial J?

Hereford→Worcester→Oxford→Reading→London

Mondays to Fridays

			2						D							125										H
			B								Ⓣ P		ⓧ	ⓧ	Ⓣ N	Ⓣ N				Ⓣ N	H		Ⓣ N			H
Hereford R	d	03 41j	06 05	—	—	—	—	—	—	—	—	—	—	—	—	—	—	—	—	—	—	—	—	—	—	—
Ledbury	d		06 21	—	—	—	—	—	—	—	—	—	—	—	—	—	—	—	—	—	—	—	—	—	—	—
Colwall	d		06 29	—	—	—	—	—	—	—	—	—	—	—	—	—	—	—	—	—	—	—	—	—	18 45	—
Great Malvern	d		06 34	—	—	—	—	—	—	—	—	—	—	—	—	—	—	—	—	—	—	—	—	—	19 05	—
Malvern Link	d		06 38	—	—	—	—	—	—	—	—	—	—	—	—	—	—	—	—	17 40x	—	—	—	—	19 13	—
Worcester Foregate Street	d		06 46	—	—	—	—	—	—	—	—	—	—	—	—	—	—	—	—	17 48x	—	—	—	—	19 19	—
Worcester Shrub Hill R	d		06 53	07 00	—	—	—	—	—	—	—	—	—	—	—	—	—	—	—	17 53x	—	—	—	—	19 22	—
Pershore	d				—	—	—	—	—	—	—	—	—	—	—	—	—	—	—	17 56x	—	—	—	—	19 33	—
Evesham	d		07 09		—	08 09	—	—	—	—	—	—	—	—	—	—	—	—	—	18 07x	—	—	—	—	19 38	20 15
Honeybourne	d		07 16		—		—	—	—	—	—	—	—	—	—	—	—	—	—	18 16	—	—	—	—	19 47	—
Moreton-in-Marsh	d		06 15 07 29		07 45	08 25	—	—	—	—	—	—	—	—	—	—	—	—	—	—	—	—	—	—	19 57	—
Kingham	d		06 23 07 38		07 53	08 34	—	—	—	—	—	—	—	—	—	—	—	—	—	—	—	—	—	—	20 04	—
Charlbury	d		06 37 07 48		08 08	08 44	—	—	—	—	—	—	—	—	—	—	—	—	—	—	—	19 20	—	—	20 20	—
Oxford	S	d	07 00 08 03		08 32	08 59	—	10 11	—	11 27	12 09	—	—	14 50	15 07	—	—	—	—	—	—	19 29	—	—	20 29	—
Didcot		d	— 08 03		08 32	08 59	—	10 11	—	11 27	12 09	—	—	14 50	15 07	—	—	—	—	—	—	19 39	—	—	20 39	—
Reading	E	S	07 17 07 21 08 54		08 48	09 42	—	10 29	11 31	11 58	12 27	13 12	13 31	—	15 03 17 25	17 57	—	—	—	—	—	19 55	21 03	—	21 33	22 18
Paddington		a	07 30 07 40 08 33 09 08	10 09 07	09 37	09 53	10 15 10 29	10 58	11 15p 11 27b 12 21	13 02 13 29	13 38 14 01	14 31w 16 07	14 38 15 07	15 25 15 54	17 53 18 05	18 30	—	—	—	—	—	20 19 20 50	21 06 21 49j	20 32r 20 48	21 35	22 17v 22 31	23 19v 23 01

Notes

A Also stops Handborough 17 40, Combe 17 43, Finstock 17 49, Ascott-under-Wychwood 18 00, Shipton 18 04
B Also stops Shipton 06 28, Ascott-under-Wychwood 06 31, Finstock 06 40, Combe 06 46, Handborough 06 49
C Also stops Shipton 07 33, Ascott-under-Wychwood 07 37, Finstock 07 47, Combe 07 52, Handborough 07 56
D Also stops Shipton 07 58, Ascott-under-Wychwood 08 02, Finstock 08 12, Combe 08 17, Handborough 08 21
E Railair links with Heathrow and Gatwick airports available to/from Reading. See separate panels for details
G Also stops Shipton 10 15, Ascott-under-Wychwood 10 18, Finstock 10 29, Combe 10 35, Handborough 10 39
H Via Swindon. Service scheduled to be operated by InterCity 125 between Paddington and Swindon and vice versa
J **Cotswold & Malvern Express**
N Via Newport. Service scheduled to be operated by InterCity 125 between Paddington and Newport
P Change at Cheltenham Spa. Service scheduled to be operated by InterCity 125 between Cheltenham and Paddington
Q Via Newport
R Seats may be reserved on certain trains from this station
S Seats may be reserved from this station through the computerised seat reservation system on certain connecting services between Paddington, Reading and Oxford, also on most services between Paddington, Reading and Swindon (note H) and between Paddington, Reading and Newport (note N), also on through trains denoted by [S] at the head of the column
a Arrival time
b Change at Oxford
c By changing at Oxford, passengers may depart at 15 22
d Departure time
e By changing at Oxford, passengers may depart at 16 42
f Change at Didcot
g By changing at Didcot, passengers may depart at 21 29
h Sunday mornings from 6 January arrive 00 19
j Tuesdays to Fridays dep 03 49
k By changing at Oxford, passengers may depart at 16 45
m Change at Swindon and Newport. InterCity 125 between Swindon and Newport only
n Change at Oxford and Didcot
p By changing at Oxford, passengers may arrive at 10 51
q Special 'bus connection
r Change at Oxford and Reading
s Special 'bus connection between Gloucester and Swindon and vice versa
t By changing at Reading, passengers may arrive at 21 35
v By changing at Oxford, passengers may arrive Reading 21 53, Paddington 22 27
w Change at Newport and Swindon
x Second class only
y Change at Cheltenham Spa
z Change at Bristol Parkway and Newport
§ Special 'bus connection between Worcester and Gloucester and vice versa
[2] Second Class only
[125] Service scheduled to be operated by InterCity 125 for all or part of journey
FO Fridays only
⊊ Drinks and cold snacks available for whole or part of journey
Ⓣ Hot dishes to order, also drinks and cold snacks available, for whole or part of journey
[S] Seats may be reserved on this train through the computer seat reservation system. Reservations may be made at London Paddington, Reading or Oxford, or at other principal British Rail Travel Centres throughout the Western Region.

British Rail car parks are provided at all stations shown in this folder except Worcester Foregate Street.

TRANSPORT USERS CONSULTATIVE COMMITTEE
If you have a complaint about British Rail services that has not been dealt with to your satisfaction, your local Transport Users Consultative Committee, set up by Parliament, will help. The address is on the T.U.C.C. notice at railway stations and is in the telephone book.

The British Railways Board accepts no liability for any inaccuracy in these tables which may be altered or cancelled at short notice, particularly during public holiday periods.

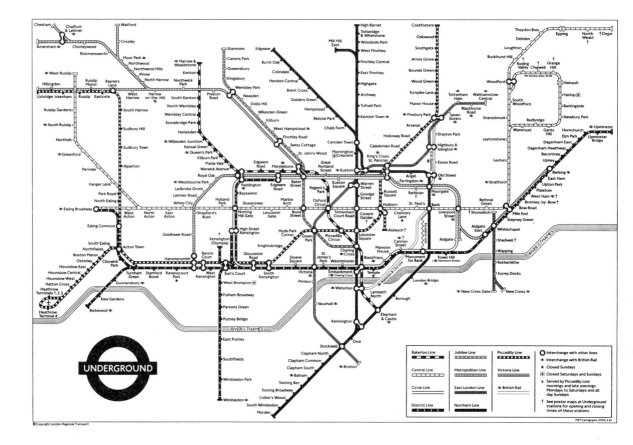

Task 2

Your train arrives in London on time, and now you have to make your way to Covent Garden. You study the map of the London Underground and realise you cannot travel there without changing trains. Assuming you only want to change trains once, answer the following questions:

1. How many stations will you go through before you change trains?
2. What is meant by an *interchange* station.
3. Name the first station you should come to.
4. At which station will you change trains?
5. What will be the destination of your new train?
6. How many more stations will you come to before you reach Covent Garden?

Task 3

Write a letter to a friend in hospital telling them about your trip to London (which was very enjoyable until something went drastically wrong!).

Task 4

The manager was delighted with the terrines. Describe a dish he could use them for.

Assignment 23: The 'open' evening

There is to be an 'Open' evening at your College, and in addition to the usual visual exhibits and practical demonstrations, the Head of the Catering Department has asked students to prepare group presentations which will be of interest to the visitors and an excellent exercise in 'communication' between the College and the general public.

Task 1

In groups of not more than five, choose a topic from the following list (or choose a topic of your own) and prepare your group presentation; each student speaking for about five minutes on his or her contribution to the project. Remember that visual aids are a great help in work of this kind.

The Kitchen Brigade
Food Service
The Bar
Coffee
English Cheeses
French Cheeses
Champagne
A Meal to Remember
The Restaurant Staff
A Wine Region

Assignment 24: How a French housewife uses chicken

Read the following passage carefully.

The rural French housewife mixes chopped fresh pork or pure pork sausage meat with eggs and herbs to stuff a big fat fowl, she poaches it with vegetables and a bouquet of herbs and the result is that *poule au pot* which good King Henry of Navarre wished that all his subjects might eat on every Sunday of the year. Or perhaps that same housewife will cook her chicken without a stuffing and serve it with a dish of rice and a cream sauce; or if it is a plump young bird, she will roast it simply in butter and serve it on the familiar long oval dish with a tuft of watercress at each end and the buttery juices in a separate sauce-boat. The farmer's wife, faced with an old hen no longer of use for laying, will (if she has inherited her grandmother's recipes and has a proper sense of the fitness of things) bone the bird, stuff it richly with pork and veal and even, perhaps, truffles if it is for a special occasion, and simmer the bird with wine and a calf's foot to make a clear and savoury jelly, so that the old hen will be turned into a fine and handsome galantine fit for celebrations and feast days.

If she is in a hurry, the French cook will cut up a roasting chicken into joints, fry them gently in butter or oil, add stock or wine, perhaps vegetables and little cubes of salt pork as well, and the result will be the *poulet sauté* which, in a restaurant, will be glorified with some classic or regional label, or named after a minister or a famous writer or actress. *Parmentier* it will be if there are little bits of potato; *provençale* if there are tomatoes; *chasseur* or *forestière* if there are mushrooms; *Poincaré* if there are asparagus tips; *Mistral* if there are aubergines; *Célestine* if there are tomatoes and wine and mushrooms and cream all together. (Célestine was one of the Emperor Louis Napoleon's cooks, and he came from the Ardéche, but the dish became celebrated at a Lyon restaurant so whether Célestine invented it or not I do not know nor, I suppose, does anybody else.)

(From *French Provincial Cooking* by Elizabeth David)

Task 1

Now answer the following questions:
1. How does the French housewife cook a 'big fat fowl'?
2. What dish did King Henry of Navarre wish his subjects might eat every Sunday?
3. What sort of chicken might be 'roasted simply in butter' and served with a garnish of watercress?
4. Which words in the passage tell you that if the French housewife bones and stuffs 'an old hen' and simmers it 'with wine and a calf's foot', she is probably using an old recipe that has been handed down to her?
5. Into what dish is the 'old hen' transformed by the French housewife?
6. If time is short, how might the French housewife deal with a roasting chicken?
7. If tomatoes were added to a basic poulet sauté, what word on the menu would describe the resultant dish?
8. What would be added to the poulet sauté to merit the addition of the word Mistral?
9. Why do you think the poulet sauté cooked with wine and mushrooms would be described as 'chasseur', or 'forestière'?
10. In which town did the dish 'Poulet Sauté Célestine' become famous?

Task 2

Using the resources of your College library and your own research, plan a menu for a 19th. century dinner.

Task 3

Using the resources of your College library and your own research, describe a dish from a particular region in France, or a country of your own choice.

Guide-lines on applying for a job

If you look at the advertisements on the opposite page, most of them ask you to 'apply in writing', and some of them ask you to 'contact' someone. This could mean making a telephone call, or writing a letter, or possibly filling in a job application form. The most important thing to remember is that your particular application will be only one of a great many, and that a badly or carelessly written, misspelt form or letter will not be very well received. So make sure that the presentation is good, and that the spelling and grammar are correct.

The letter

This should contain the following information: age, status (married or single), names of secondary schools and colleges attended. (No further details are needed in the letter itself – you should write these on a separate sheet of paper and attach it to the letter.) Include details of your experience relevant to the job, and something about yourself – your interests or hobbies – which would help a future possible employer to learn something about you as a person. The whole object of writing the letter is to make a possible employer want to interview you. So if you want to get a job, get the application right.

The form

This should be completed in pencil first to make quite sure that you haven't made any mistakes. Read instructions carefully. Where you are told to use 'block capitals' – use them. Where you are asked to fill in details of your interests or hobbies be careful what you write. One student of mine when filling in an application form for a job as a typist wrote that she was 'interested in all-night discos and riding pillion on her boyfriend's motor bike'. This comment might not fill a future employer with confidence!

The telephone call

The notes on 'Using the Telephone', pp. 2 and 3, apply equally here. Be *confident* – at this stage you have everything to gain and nothing to lose. Be *polite* – sometimes shyness can make people sound rude. Be *positive* – remember that you are ringing up to *inquire* about the job and then take it from there.

The C.V. (Curriculum Vitae)

This should be presented on a separate sheet of paper to be enclosed with a letter of application. It should contain full details of your schools, (other than primary schools), and colleges and details of examinations taken and their results. (Only give these when you have been successful, there is no point in giving details of examinations you have failed!) You should also give full details of the contributions you have made to your school or college. For example — did you take part in any plays or productions; were you a member of any games team?; or did you play for the school band or orchestra? Did you take a prominent part in any fund-raising activity? Do not under-present yourself! All this information should be neatly tabulated so that it can be very easily read and understood, under appropriate headings.

Task 1

Write your own C.V.

Task 2

Look at the advertisements taken from the *Caterer and Hotel Keeper*. Choose one of them and prepare a letter of application for the job.

Task 3

See if you can get some job application forms and practise filling them in. (Your local Job Centre may be able to help you.)

EYRIE HOTEL

As one of Scotland's foremost hotels, Eyrie is a byword for elegance and excellence.

To maintain our prestigious position we are continuing to improve the standard of service.

This winter sees the addition of 53 new luxury bedrooms to enable us to cater for the all year round demand for the hotel. Recruiting the right staff is just part of our Investments for the future.

ALL GRADES OF STAFF

We are looking for experienced staff at all levels of the organisation, particularly for Spring 1986, to cope with the increased levels of business we are achieving throughout the year.

To the right people we can offer first class conditions including accommodation if required, meals, generous holiday allowances and attractive pension and sick pay schemes, in fact all the benefits you'd associate with working with a dynamic and progressive hotel group. Write to Mr. Cameron, Personnel Manager, Eyrie Hotel, Kincraig, Perthshire PH6 1NZ

EYRIE HOTEL

FLOOR HOUSEKEEPERS

Join our young efficient team. We are a large modern hotel situated just off Kensington High Street and are looking for experienced Floor Housekeepers capable of supervising 100 rooms. You will work a five-day week from 7.30am-4pm. We are offering up to £98 pw + bonus and other fringe benefits.

Please contact:
Personnel Department, Astra Hotel,
Wrights Lane, Kensington, London W8
Tel: 832 7811 between 9am-5pm

HVCA2-182

ASSISTANT MANAGER
Required

Due to increase in business to assist in all aspects of 43 bedroomed hotel. May suit college leaver or trainee as first position. Salary by negotiation. Live-in position.

Apply in writing with C.V. to date to:
Mr. Johnston, Trafford Motor Inn Ltd.,
Coventry Rd, Solihull.

RECEPTIONIST

Required for a busy hotel situated in Leicester Square.

Experience of NCR 42 billing machine and PMBX switchboard required.

Applications to:
The Manager • The Astoria Hotel • 23 St Martin's Lane
Leicester Square, London WC2
Tel: 01 930 2750

FVCS2-121

IMMEDIATELY REQUIRED
A YOUNG AMBITIOUS

SECOND CHEF

for busy country inn on the Kent/Sussex borders. 30-seater restaurant, must be able to handle à la carte, bar meals and motivate staff. Live in or out. Good salary.

Apply:
Mrs. G. Andrewes, The Wheatsheaf, Newenden, Kent.

HEAD CHAMBERPERSON

Required for *remote* 20-bedroomed Country Hotel in North West Highlands of Scotland. Live-in. May suit couple as we also require BAR PERSON AND SECOND CHEF

Please write, with full C.V., to:
G. Hartley, Queen's Hotel, Glencoe, Argyll, Scotland

Under the direction of
MONSIEUR CHARLES GIDE
of Les Marroniers Restaurant, Oxford
Positions are available due to expansion of the restaurant:

Les Marroniers, Burford, Oxford ★★ Michelin

CHEF DE RANG (M/F)
COMMIS DE RANG (M/F)
CHEF DE PARTIE (M/F)

All applications in writing sending CV and photograph to:
Les Marroniers, Burford, Oxford

RV

Manor Hotel
Stratford-Upon-Avon requires

WAITING & BAR STAFF

This highly rated 27 bedroom Elizabethan Manor House urgently requires restaurant and bar staff.

The successful applicants must have flair and the experience to deal with our very discerning clientéle. Accommodation available if required.

Apply in writing with full C.V. and photograph to:-
Restaurant Manager, Manor Hotel, Stratford-Upon-Avon, Warwickshire.

RESTAURANT MANAGER
and
ASSISTANT MANAGER

We are an expanding 80 bedroomed Georgian Manor Hotel with a thriving Conference, Banqueting and à la carte trade.

The ideal applicants will have the personality and ability to contribute towards the continuing success of this privately owned hotel with extensive sports facilities including a 9 hole golf course.

Commencing salary for the Restaurant Manager will be circa £7,500 pa and the Assistant Manager will be £6,500 pa. Excellent live-in accommodation available.

For informal discussion please contact:
Helen Anderson, Avalon Manor, Colchester, Essex.

Guide-lines on the interview

You have applied for a job and have been selected for an interview!

This is one of the occasions when it is very important to present yourself well. So, bear in mind the following points:

1. Remember the job you're hoping to get and wear clothes and a hair-style that is appropriate for it.
2. When you meet the person who is going to interview you greet them with a smile and a handshake (which should not be of the 'rugby player crunch' or the 'limp lettuce' variety!). The smile will help to make you feel more relaxed if you're feeling nervous, because your facial muscles will lose their tenseness.
3. When you are asked to sit down, try to sit comfortably in your chair without slumping or sagging in it. The minute you slump you lose authority.
4. Do answer questions as fully as possible. You may be feeling nervous and shy but that is no excuse to answer with a mere 'yes' and 'no'. Remember you are there to get a job and you must contribute as much as possible to the interview. Be courteous and pleasant at all times – never 'off-hand' which can be interpreted as rudeness, even if that wasn't the intention. Finally do not underestimate yourself. Be *confident* and *enthusiastic* and you will impress.

Task 1

List all the things you would want to know at an interview.

Task 2

Most candidates are invited at some time or other by the interviewer to 'tell me something about yourself'. In pairs, spend a short time finding out each other's hobbies, likes and dislikes and then report your discoveries to the rest of the group.